Scientific and Technological Flows Between the United States and China

JON SCHMID, NATHANIEL EDENFIELD

Prepared for the Office of the Secretary of Defense
Approved for public release; distribution unlimited

NATIONAL DEFENSE RESEARCH INSTITUTE

For more information on this publication, visit **www.rand.org/t/RRA2308-1**.

About RAND

The RAND Corporation is a research organization that develops solutions to public policy challenges to help make communities throughout the world safer and more secure, healthier and more prosperous. RAND is nonprofit, nonpartisan, and committed to the public interest. To learn more about RAND, visit www.rand.org.

Research Integrity

Our mission to help improve policy and decisionmaking through research and analysis is enabled through our core values of quality and objectivity and our unwavering commitment to the highest level of integrity and ethical behavior. To help ensure our research and analysis are rigorous, objective, and nonpartisan, we subject our research publications to a robust and exacting quality-assurance process; avoid both the appearance and reality of financial and other conflicts of interest through staff training, project screening, and a policy of mandatory disclosure; and pursue transparency in our research engagements through our commitment to the open publication of our research findings and recommendations, disclosure of the source of funding of published research, and policies to ensure intellectual independence. For more information, visit www.rand.org/about/principles.

RAND's publications do not necessarily reflect the opinions of its research clients and sponsors.

Published by the RAND Corporation, Santa Monica, Calif.
© 2023 RAND Corporation
RAND® is a registered trademark.

Library of Congress Cataloging-in-Publication Data is available for this publication.

ISBN: 978-1-9774-1134-1

Cover images: Livinskiy Ivan Evgenyevich/Adobe Stock; Elen11/Getty Images

About This Report

Scientific and technological competition has emerged as a front on which strategic competition between the United States and China is contested. Scientific and technological dominance—the prize of this competition—has been recognized as a national priority by high-level leadership from both countries. This dominance can be attained in two primary ways: A country can rely on its domestic scientific and technology innovation resources and activity, or it can leverage foreign scientific and technological assets. This study focuses on the second approach. Specifically, this study will investigate three types of flows between the United States and China: the inflow of U.S. technology inputs into Chinese military technology, the bilateral movement of scientific researchers between the United States and China, and scientific collaboration between U.S.- and China-based researchers. The research reported here was completed in January 2023 and underwent security review with the sponsor and the Defense Office of Prepublication and Security Review before public release.

RAND National Security Research Division

This research was sponsored by the Office of the Secretary of Defense and conducted within the Acquisition and Technology Policy Program of the RAND National Security Research Division (NSRD), which operates the RAND National Defense Research Institute (NDRI), a federally funded research and development center sponsored by the Office of the Secretary of Defense, the Joint Staff, the Unified Combatant Commands, the Navy, the Marine Corps, the defense agencies, and the defense intelligence enterprise. For more information on the RAND Acquisition and Technology Policy Program, see www.rand.org/nsrd/atp or contact the director (contact information is provided on the webpage).

Acknowledgments

The authors would like to thank Chad Ohlandt, Chad Heitzenrater, Joel Predd, Yun Kang, Christopher Mouton, Michael Kennedy, Christian Curriden, and Marjory Blumenthal for their insightful comments on previous versions of the research presented here.

Summary

Issue

Scientific and technological competition has emerged as a front on which strategic competition between the United States and China is contested. Scientific and technological dominance—the prize of this competition—has been recognized as a national priority by high-level leadership from both countries. Strategic competition colors how international science and technology flows between the United States and China are interpreted in the United States. On one hand, the international movement of researchers and knowledge promotes an efficient allocation of scientific and technological resources, increasing the productivity of the U.S. knowledge production system. On the other hand, international flows might present national security risks if knowledge and technology developed in the United States are used by China to modernize its military or otherwise gain competitive advantage. However, before the net effect of international science and technology flows can be assessed, these flows must first be measured and described. The objective of this report is to quantify and describe three types of international flows and to document the empirical methods used to do so.

Approach

This report considers three distinct types of international flows using three distinct empirical approaches and data sources. To assess the technological inputs to patented Chinese military technology, we perform citation analysis of the prior art listed in Chinese military patents.[1] To assess the movement of researchers between the United States and China over time, we track changes in authors' affiliations on scientific publications. Finally, to characterize international collaboration on aerospace engineering research, we use coauthorship data from scientific publications.

Findings

Findings on the Inputs to Chinese Military Technology

- Most prior art cited in Chinese military patents comes from other Chinese organizations.

---

[1] *Prior art* refers to any material (in practice, typically other patents or non-patent literature, such as scientific publications) that can establish that the invention under application satisfies the patentability criteria of novelty and nonobviousness.

- The United States is the largest foreign source of prior art in patented Chinese military technology; technologies developed by the U.S. Army and U.S. Navy are particularly highly utilized.

Findings on International Researcher Flows

- Under 3 percent of researchers in our sample who were based in the United States or in China were internationally mobile (i.e., obtained an affiliation with a university in the other country) between 2011 and 2020.
- Of the internationally mobile researchers in our sample, U.S.-based researchers migrated to a Chinese affiliation more often than leaving and returning to the United States. China-based researchers more often returned to China after a period in the United States.
- Most internationally mobile researchers obtained multi-affiliations at some point. Researchers who obtained a multi-affiliation returned to an affiliation in their home country or maintained their multi-affiliation more often than they ultimately migrated.
- Among internationally mobile researchers, returnees had the greatest influence via number of citations.

Findings on U.S.-Chinese Research Collaboration

- The effect of U.S.-Chinese research collaboration should be viewed in terms of threats and benefits. We find evidence of both.
  - In terms of threats, U.S.-based organizations have coauthored more aerospace engineering publications with Chinese organizations with linkages to the People's Liberation Army in recent years than in the past.
    - There is a nontrivial amount of U.S.-Chinese research collaboration on sensitive topics, such as hypersonic systems.
  - In terms of benefits, we find that aerospace publications written by teams composed of researchers from the United States and China have greater-than-average influence and are more interdisciplinary.

Meta-Finding

- The character and magnitude of international science and technology flows can be assessed by mining patent and publication data. However, sound policy will depend on careful assessment of the costs and benefits to openness.

Contents

Figures and Tables

Figures

Tables

CHAPTER 1

Introduction

Scientific and technological competition has emerged as a contested front for strategic competition between the United States and China. Scientific and technological dominance—the prize of this competition—has been recognized as a national priority by high-level leadership from both countries.[1] This dominance can be attained in two nonmutually exclusive ways. A country can rely on its *domestic* scientific and technology innovation resources and activities, or it can leverage *foreign* scientific and technological assets.[2] This report focuses on the second approach. Specifically, we will investigate three types of flows between the United States and China: the inflow of U.S. technology inputs into Chinese military technology, the bilateral movement of scientific researchers between the United States and China, and scientific collaboration between researchers based in the United States and researchers based in China.

The intention of these three investigations is relatively modest. We seek to describe the character of these flows in recent years and to develop empirical methods to do so. No policy or normative conclusions are drawn. We do not offer policy recommendations in this report because the effects of the international movement of scientific and technological flows on national security are often ambiguous.

Consider a Chinese scientist who completes a doctorate degree at Stanford University and a two-year postdoctoral program at a Department of Energy laboratory and then returns to China via a talent program to conduct research at the National University of Defense Technology. Perhaps the primary potential harm to U.S. national security associated with this type of movement is that the scientist in question will contribute to Chinese military modernization via the knowledge acquired while a graduate student and postdoc in the United States. However, this potential harm must be weighed against the benefit to U.S. national

---

[1] James Kynge, "China's High-Tech Rise Sharpens Rivalry with the US," *Financial Times*, January 19, 2022; Graham Allison, Kevin Klyman, Karina Barbesino, and Hugo Yen, "Avoiding Great Power War Project: The Great Tech Rivalry: China vs the US," Harvard Kennedy School Belfer Center for International Affairs, 2021, p. 73.

[2] Popper et al. provide a conceptual framework, along with a measurement approach, for assessing China's innovation potential and its likelihood of realizing that potential (Steven W. Popper, Marjory S. Blumenthal, Eugeniu Han, Sale Lilly, Lyle J. Morris, Caroline S. Wagner, Christopher A. Eusebi, Brian Carlson, and Alice Shih, *China's Propensity for Innovation in the 21st Century: Identifying Indicators of Future Outcomes*, RAND Corporation, RR-A208-1, 2020).

security associated with the student's positive contribution to the U.S. scientific enterprise during their time in the United States or collaboration thereafter.[3] This report focuses on empirical characterization of international flows; we recognize these potential harms and benefits but make no judgement here about the net effect of these flows on national security. We believe, however, that assessment of the net effect of scientific and technological openness and the likely effects of various policy options is the essential next step to this research.

While we make no normative judgements regarding the effect of the flows assessed here on national security, it is worth noting that the Chinese Communist Party (CCP) and the People's Liberation Army (PLA) are actively seeking to promote a flow of U.S. scientific and technological assets into China. China runs hundreds of talents programs that provide incentives to attract highly skilled individuals to work in China. Some of these programs explicitly seek to advance China's military modernization priorities.[4] Additionally, the CCP's "Going Out" strategy attempts to advance technology priorities by acquiring foreign firms, establishing labs and technology listening posts abroad, and attracting overseas talent. For example, China uses science and technology (S&T) diplomats and professional associations to identify international technology investment opportunities (often of military relevance) for Chinese firms and investors.[5] In the analyses to follow, we pay particular attention to the flows to and from organizations with strong ties to the PLA because these flows might be of particular interest to national security scholars and policymakers, given China's efforts to encourage S&T assets inflow from the United States.

The three analyses that follow are exploratory in nature. Each investigates a different flow using a distinct dataset and methodology. Chapter 2 uses patent data to determine the organizations and countries that contribute the most prior art to Chinese military patents. Chapter 3 uses publication data from Scopus to track researcher affiliation changes between a sample of prominent universities in the United States and China while also tracking researcher impact and focus.[6] Chapter 4 explores the recent increase of academic collaboration between the United States and China in aerospace engineering, using Web of Science publication data, to explore the potential sources of risk and benefit to the United States.[7] There are similarities between the analyses. For example, each tracks knowledge flows down to the organization

---

[3] Attendant harms and benefits to national security international flows can also accrue to China. This example takes the perspective of the United States to illustrate the ambivalence of these effects. Furthermore, the economic benefits of scientific openness can have national security effects for either country.

[4] Rob Portman and T. Carper, *Threats to the U.S. Research Enterprise: China's Talent Recruitment Plans*, staff report to the Permanent Subcommittee on Investigations, U.S. Senate, November 18, 2019.

[5] Ryan Fedasiuk and Emily Weinstein, "Overseas Professionals and Technology Transfer to China," *Issue Brief*, Center for Security and Emerging Technology, July 21, 2020; Ryan Fedasiuk, Emily Weinstein, and Anna Puglisi, *China's Foreign Technology Wish List*, brief, Center for Security and Emerging Technology, May 2021.

[6] Scopus is Elsevier's abstract and citation database. For more information about the author affiliations cited in this report, see Elsevier, "Scopus: Affiliations," webpage, undated-b.

[7] Clarivate, "Web of Science: Search," webpage, undated-b.

or university level: an often-overlooked approach to studying international flows between the United States and China. At the same time, the results of these analyses do not merge to answer a single research question. Instead, they provide the reader with an empirical characterization of the flow in question and with a description of the methodologies used to do so.

CHAPTER 2

Identifying the Sources of Technological Inputs to Chinese Military Technology

Chinese military technology—like all technology—builds on ideas, techniques, and technologies from the past. The objective of this chapter is to describe the sources of the inputs to Chinese military technology by examining the citations contained within patented Chinese military technology. We paid particular attention to inputs sourced from the United States (i.e., prior art from U.S. inventors). By analyzing the patents that are cited by Chinese military patents, we sought to answer the following three questions: (1) What countries contribute the most prior art to patented Chinese military technology? (2) What organizations contribute the most prior art to patented Chinese military technology? (3) What types of patents are being cited by Chinese military patents?

Data

The data used here are drawn from patents in the Derwent Innovation Index (DII), a database containing patent grants from all major national (e.g., the U.S. Patent Office and the Chinese Patent Office) and international (e.g., the European Patent Office) patent agencies.[1] In the following analysis, we use a dataset of the 2,728 military patents[2] that were granted to

---

[1] Clarivate, "Derwent Innovations Index on Web of Science," database, undated-a.

[2] *Military patents* refers to patents assigned the W07 (Electrical Military Equipment and Weapons) Derwent Classification Code. Schmid shows that this class of patents adhere to both a scholarly and common-sense definitions of new military technology (Jon Schmid, "The Determinants of Military Technology Innovation and Diffusion," dissertation, Georgia Institute of Technology, 2018). However, patents are an imperfect measure of military technology innovation. The intellectual properties underlying many military technologies are maintained as trade secrets. China maintains certain patents as secret for national security reasons. In other instances, military technology can be attained illicitly via intellectual property theft. Trade secrets, secret patents, and intellectual property theft would not be captured in the patent dataset used here. Patent and publication data also have lags (e.g., the science underlying a publication, by necessity, takes place prior to the publication that documents the scientific process and results), which complicates teasing out the effect of particular policy changes. Finally, China previously gave incentives to patent, which led to lower-than-average patent quality (Jon Schmid and Fei-Ling Wang, "Beyond National Inno-

Chinese organizations from 2016 to 2020.[3] *Chinese organizations* refers to organizations that are based in China.[4]

Figure 2.1 depicts one of the Chinese military patents (CN-105841556-B) used in this investigation. The patent is for a "general advanced upper stage of solid launch vehicle" and was granted by the Chinese Patent Office on April 12, 2017.[5] The patent was granted to the General Designing Institute of Hubei Space Technology Academy, a research organization located within Hubei University of Technology. Chinese-language documents, such as that shown in Figure 2.1, are machine-translated by Clarivate, the owners of the DII patent database.

Patent Citations

Patent applicants are required to cite as prior art all patented technologies relevant to the invention under application. These citations (known as *backward citations*) therefore constitute a set of technological inputs into the focal technology.[6] The sources of these inputs are the focus of this chapter.

Figure 2.2 illustrates the patent citation process. The patent described above (CN-105841556-B) cites eight patents as prior art. Three of these were U.S. patents (i.e., patents granted to U.S. organizations) and the other five were Chinese patents. In consideration of space, only four of these patents are depicted in Figure 2.2.

---

vation Systems: Incentives and China's Innovation Performance," *Journal of Contemporary China*, Vol. 26, No. 104, 2017).

[3] To determine the year of each patent, we use the patent's *priority year* (the year during which the patent application was first filed). This year constitutes the closest date to when the underlying invention was conceived and using it thus reduces the effect of the patent-application/patent-grant time lag.

[4] For the vast majority of patenting organizations, this means that virtually all of their internal business operations take place within China. For multinational corporations (e.g., Tencent), we determine nationality based in on the location of the corporate headquarters.

[5] Google Patents, "General Advanced Upper Stage of Solid Launch Vehicle," webpage, undated.

[6] *Forward citations* refers to the citations received by a given patent by subsequent patenting.

FIGURE 2.1

Example of the Front Page of a Chinese Military Patent

SOURCE: Reproduced from Google Patents, undated.

FIGURE 2.2

The Patent Citation Process

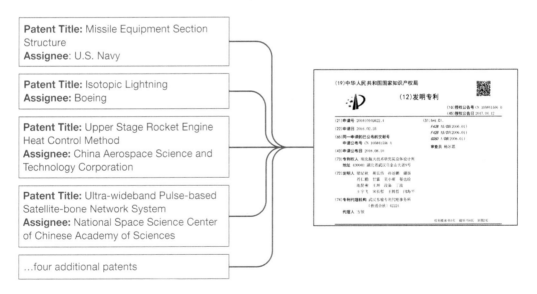

SOURCES: Left side features DII data (Clarivate, undated-a), and right side is reproduced from Google Patents, undated.

Results

What Countries Contribute the Most Prior Art to Patented Chinese Military Technology?

Perhaps unsurprisingly, most of the prior art cited by Chinese military technology comes from China. Chinese organizations developed 2,211 (87.6 percent) of the patents that were cited by Chinese military patents. Table 2.1 provides the country-level contributions of patents that were cited by Chinese military patents. U.S. patents were cited 162 times (6.4 percent) by Chinese military patents.

What Organizations Contribute the Most Prior Art to Patented Chinese Military Technology?

Table 2.2 depicts the organizations cited most frequently by Chinese military patents. Chinese organizations occupy eight of the top ten most-cited organizations, with seven of the eight being universities. The large contribution made by Chinese universities to prior art in Chinese military technology is atypical; the primary innovative contributions of other countries tend to come from government and commercial entities. This is consistent with other research that finds that Chinese universities are highly active in patenting and the commercialization of invention.[7] It is worth noting that the U.S. Navy and the U.S. Army are the

TABLE 2.1

Country-Level Contribution to Chinese Military Patent Citations

Country	Patents Cited by 2016–2020 Chinese Military Patents	Share of Total (%)
China	2,211	87.6
United States	162	6.4
Japan	67	2.7
Germany	41	1.6
South Korea	30	1.2
Other	14	0.6
Total cited patents	2,525	

SOURCE: Features DII data (Clarivate, undated-a).

---

[7] Schmid and Wang, 2017; Hong Gong, Libing Nie, Yuyao Peng, Shan Peng, and Yushan Liu, "The Innovation Value Chain of Patents: Breakthrough in the Patent Commercialization Trap in Chinese Universities," *PLoS ONE*, Vol. 15, No. 3, 2020.

TABLE 2.2

Organization-Level Contribution to Chinese Military Patent Citations

Assignee	Patents Cited by 2016–2020 Chinese Military Patents
Beijing Institute of Technology	179
Beihang University	144
China Academy Launch Vehicle Technology	109
Harbin Institute of Technology	97
Nanjing University of Science and Technology	90
North University of China	57
Ordnance Engineering College	45
Northwestern Polytechnical University	38
U.S. Navy	37
U.S. Army	36

SOURCE: Features DII data (Clarivate, undated-a).

ninth and tenth most frequently cited organizations. That is, patents developed by the U.S. Navy and the U.S. Army are frequently cited as prior art in Chinese military patents.

Figure 2.3 depicts the largest organization- and country-level contributors of prior art to Chinese military patents. Figure 2.3 visually demonstrates, as Table 2.1 and 2.3 numerically show, that Chinese organizations are the largest contributors of prior art to patented Chinese military technology.

Table 2.3 displays the top non-Chinese organizations according to how often they were cited as prior art by Chinese military patents. These organizations represent a mix of defense-focused organizations (e.g., U.S. Army, U.S. Navy, and Agency for Defense Development) and dual-servicing (i.e., military and civilian) organizations (e.g., Boeing, Mitsubishi, and Toyota). In contrast to the top Chinese organization-level contributors, none of these organizations are universities.

Figure 2.4 displays the largest non-Chinese organization- and country-level contributors to the prior art of Chinese military technology. The United States is the largest non-Chinese contributor to Chinese military technology prior art, contributing 52 percent of the non-Chinese citations found in the Chinese military technology patents. Japan, Germany, and South Korea are the next largest country-level contributors, contributing 20 percent, 14 percent, and 10 percent of the non-Chinese prior art in the Chinese military technology patents studied here.

FIGURE 2.3

Prior Art Flows into Chinese Military Patents—All Sources

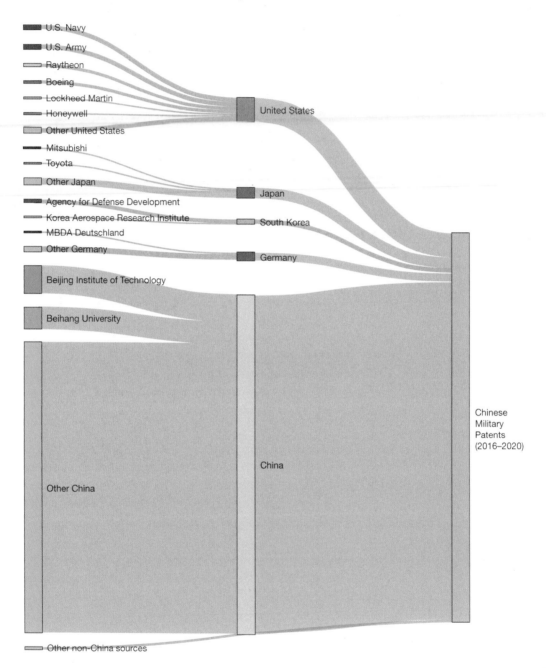

SOURCE: Features DII data (Clarivate, undated-a).

TABLE 2.3

Top Non-Chinese Organization-Level Contributions to Chinese Military Patent Citations

Assignee	Host Country	Patents Cited by 2016–2020 Chinese Military Patents
U.S. Navy	United States	37
U.S. Army	United States	36
Agency for Defense Development	South Korea	22
Raytheon	United States	20
Boeing	United States	19
Mitsubishi	Japan	11
Toyota	Japan	10
Korea Aerospace Research Institute	South Korea	8
Lockheed Martin	United States	8
Lfk Lenkflugkoerpersysteme (MBDA)	Germany	8
Honeywell	United States	8

SOURCE: RAND assessment using DII data (Clarivate, undated-a).

FIGURE 2.4

Patent Flows into Chinese Military Patents—Non-China Sources

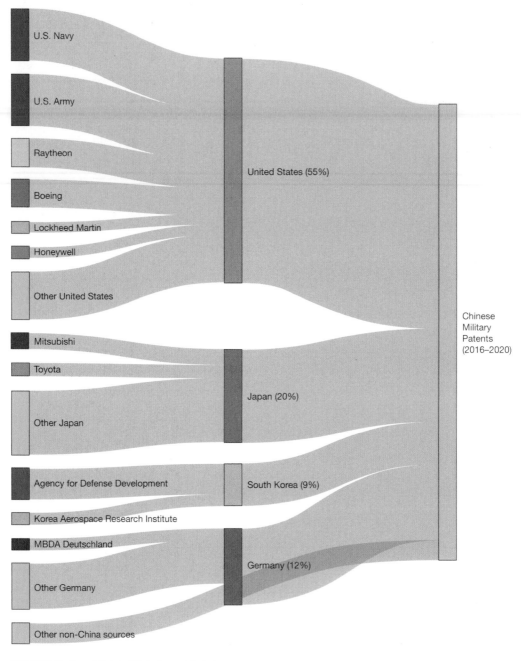

SOURCE: Features DII data (Clarivate, undated-a).

What Types of Patents Are Being Cited by Chinese Military Patents?

Table 2.4 depicts the individual patents that were cited most frequently by Chinse military patents. There were only three patents cited more than two times; the vast majority of patents cited by Chinese military patents were cited just once.

Analysis of the technical content of the cited patents reveals that Chinese military patents depend on both military and civilian technology inputs. Of the U.S. patents cited by Chinese military patents, 33 percent were for civilian technologies and 67 percent were for military technologies.[8]

Conclusions

At least three conclusions can be drawn from this investigation. First, the vast majority (87.6 percent) of citations made by Chinese military patents are for other Chinese patents. Second, the largest foreign source of citations is the United States; and technologies developed by the U.S. Army and U.S. Navy are particularly highly utilized. Finally, this investigation demonstrates that the citation data of military patents offer a novel way of investigating the sources of military technology innovation.[9]

TABLE 2.4

Most Frequently Cited U.S. Patents

Patent Title	Patent Numbers	Times Cited by Chinese Military Patents	Assignee	Application Year
Nonlethal Waterborne Threat Deterrent and Immobilization Device	US8714070B2	8	Engineering Science Analysis Corp	2013
An Over-Voltage Detection Test Apparatus for Military Aircraft Weapon Systems	US4996520A	6	Williams Instruments Inc	1988
Warhead Initiation Circuit	US4934268	3	U.S. Army	1989

SOURCE: Features DII data (Clarivate, undated-a).

---

[8] To arrive at this figure, military technologies were defined as those assigned Derwent Classification Code "W07" (Electrical Military Equipment and Weapons).

[9] In addition to the citation behavior, the content of miliary patents (i.e., the text describing the underlying invention) is a rich source of information about the character of national military technology innovation effort. Schmid demonstrates how patent text can be mined to glean insight into the specific technical focus of overall military technology innovation trends (Jon Schmid, "Technological Emergence and Military Technology Innovation," *Defence and Peace Economics*, June 2022).

CHAPTER 3

Assessing the Flow of Academic Researchers Between the United States and China

Academic researchers moving between U.S. and Chinese universities have recently come under increased scrutiny.[1] On May 28, 2020, in an effort to reduce the flow of "non-traditional information collectors" between the two countries, President Donald Trump issued a proclamation that allowed the federal government to deny or revoke visas of Chinese graduate students and researchers at U.S. universities if they have or had an affiliation with institutions that "implement or support" China's military-civil fusion strategy.[2] One source estimated that the policy will deny 3,000 to 5,000 Chinese graduate students from entering the United States each year.[3] The policy's effect on current researchers remains uncertain because of a lack of established data on international researcher flows. In turn, recent studies have begun to assess the stock of Chinese researchers in the United States and

---

[1] Recent U.S. government policy actions to counter nontraditional collection and influence activities from China include the following: Public Law 115-232, John S. McCain National Defense Authorization Act for Fiscal Year 2019; Section 1286, Initiative to Support Protection of National Security Academic Researchers from Undue Influence and Other Security Threats, August 13, 2018; U.S. Department of Energy Order 486.1, "Department of Energy Foreign Government Talent Recruitment Programs," June 7, 2019; U.S. Department of Energy Order 142.3A, "Unclassified Foreign Visits and Assignments Program," October 14, 2010; U.S. Department of Energy Order 142.3B, "Unclassified Foreign National Access Program," January 15, 2021; Public Law 116-283, William M. (Mac) Thornberry National Defense Authorization Act for Fiscal Year 2021; Section 223, Disclosure of Funding Sources in Applications for Federal Research and Development Awards, January 1, 2021; U.S. Department of Justice, "Information about the Department of Justice's China Initiative and a Compilation of China-Related Prosecutions Since 2018," November 19, 2021; White House, "Presidential Memorandum on United States Government-Supported Research and Development National Security Policy: National Security Presidential Memorandum – 33," January 14, 2021.

[2] White House "Suspension of Entry as Nonimmigrants of Certain Students and Researchers from the People's Republic of China," Proclamation 10043, *Federal Register*, Vol. 85, No. 34353, May 29, 2020.

[3] Remco Zwetsloot, Emily Weinstein, and Ryan Fedasiuk, *Assessing the Scope of U.S. Visa Restrictions on Chinese Students*, Center for Security and Emerging Technology, February 2021.

vice versa.[4] The objective of this investigation is to quantitatively assess the international movement of academic researchers between subsets of prominent U.S. and Chinese universities. We also perform a qualitative assessment of research impact and research topics for the set of researchers identified as internationally mobile between the two countries.

Data

The use of publication data to track the international movement of scientists was introduced in 2003 using a small sample of elite researchers in a specific discipline.[5] This approach has since expanded to allow tracking millions of international researchers across all disciplines with the adoption of author disambiguation techniques by publication databases.[6] We selected Scopus, a publication database containing more than 77 million records, as the source of publication data for this investigation to leverage its internally assigned Author Identifier.[7] The Scopus Author Identifier is a unique number that is algorithmically assigned to each author based on an author's affiliations, subject areas, coauthors, and other data fields recorded within Scopus. Researchers have shown Scopus author profiles to have average precision of 98.1 percent and an average recall of 94.4 percent.[8] Furthermore, research has shown Scopus publication data as a viable means of tracking international mobility among researchers.[9]

---

[4] A study published in 2020 assessed the stock of overseas and returnee Chinese scientists in the United States and Europe for 2010 and 2017, along with their research impact through citation and international collaboration. For more information, see Cong Cao, Jeroen Baas, Caroline S. Wagner, and Koen Jonkers, "Returning Scientists and the Emergence of China's Science System," *Science and Public Policy*, Vol. 47, No. 2, April 2020, pp. 172–183.

[5] The first widely cited publication to use bibliometric data to assess international mobility tracked 131 researchers. See Grit Laudel, "Studying the Brain Drain: Can Bibliometric Methods Help?" *Scientometrics*, Vol. 57, No. 2, 2003.

[6] More recently, a 2020 study used publication data to assess the global prevalence of multiple affiliations across 15 million authors from 40 countries using 22 million articles. See Hanna Hottenrott, Michael E. Rose, and Cornella Lawson, "The Rise of Multiple Institutional Affiliations In Academia," *Journal of the Association for Information Science and Technology*, Vol. 72, No. 8, August 2021.

[7] Elsevier, "Scopus: Search for an Author Profile," webpage, undated-c.

[8] *Precision* refers to the percentage of records correctly assigned to the researcher. *Recall* refers to the percentage of a researcher's records assigned to the researcher. See Jeroen Baas, Michiel Schotten, Andrew Plume, Grégoire Côté, and Reza Karimi, "Scopus as a Curated, High-Quality Bibliometric Data Source for Academic Research in Quantitative Science Studies," *Quantitative Science Studies*, Vol. 1, No. 1, Winter 2020.

[9] We note that there are still limitations to using Scopus to track international mobility. Author Identifiers for researchers with common first and last names are known to have a higher recall and lower precision. Given that common names are prevalent in Asia, a Scopus author profile of an Asian author is more likely to contain records of other authors than that of a European author profile. Asian authors thus have a higher probable migration rate. We also note that the Scopus algorithm may also split author profiles, particularly

To build our dataset of U.S. and Chinese researchers from Scopus, we collected publication data for research articles indexed under the "Physical Sciences" Scopus subject area for the period 2011 to 2020.[10] We consider researcher affiliations within one subset of U.S. universities and two subsets of Chinese universities: the Top-10 ranked U.S. and Chinese research universities (subsequently referred to as US10 and PRC10) and the Seven Sons of National Defense (Seven Sons or SSons).[11] Although an ideal investigation would consider authors from all U.S. and Chinese universities, we did not have access to this scope of data. Instead, we selected a sample of universities to provide an initial exploration of the data and our methodology. Future research could explore the movement of authors between universities of different tiers, especially given that it might be easier for international researchers to find appointments at lower-ranked universities within each country.

The top ten universities of each country contain many leading researchers within their respective fields and were selected because of their important role in advancing science in both countries. The Seven Sons universities contain researchers likely to be tied to China's military-civil fusion strategy; these universities consider themselves to be "defense science, technology and industry work units" and instruments of the "defense system."[12] A Chinese or People's Republic of China (PRC) affiliation in this chapter refers to an affiliation with any university located in China. A U.S. affiliation refers to an affiliation with any university located in the United States.

With the data assembled, we leveraged Scopus's Author Identifier and the author affiliation listed on each publication to track an author's affiliation history over time. We removed

---

in fields with many publications, lowering the probable migration rate for authors in those fields. We present mitigations later in this chapter. See Henk F. Moed, M'hamed Aisati, and Andrew Plume, "Studying Scientific Migration in Scopus," *Scientometrics*, Vol. 94, No. 3, 2013, pp. 929–942.

[10] Physical Sciences subject areas in Scopus include Chemical Engineering, Chemistry, Computer Science, Earth Science, Energy, Environmental Science, Materials Science, Mathematics, and Physics. For more information, see Elsevier, "Physical Sciences and Engineering," webpage, undated-a.

[11] The selections for the US10 were based on the 2022 Shanghai World Rankings. The universities included were California Institute of Technology, Columbia University, Cornell University, Harvard University, Massachusetts Institute of Technology (MIT), Princeton University, Stanford University, University of California (UC) Berkeley, University of Chicago, and Yale University. The selections for the PRC10 were based on the 2022 Shanghai World Rankings. The universities included were Fudan University, Huazhong University of Science and Technology, Nanjing University, Peking University, Shanghai Jiao Tong University, Sun Yat-sen University, Tsinghua University, University of Chinese Academy of Sciences, University of Science and Technology of China, and Zhejiang University. The Chinese Academy of Sciences (CAS)—including the University of the Chinese Academy of Sciences—was added because of its prominence in other rankings and the ambiguity of the ninth and tenth spots in the Shanghai World Rankings. For more information, see Shanghai Ranking, "2022 Academic Ranking of World Universities," webpage, 2022. The Seven Sons universities are Beijing Institute of Technology (BIT), Beijing University of Astronautics and Aeronautics, Harbin Engineering University, Harbin Institute of Technology (HIT), Nanjing University of Astronautics and Aeronautics, Nanjing University of Science and Technology, and Northwest Polytechnical University.

[12] Alex Joske, *The China Defence Universities Tracker: Exploring the Military and Security Links of China's Universities*, Australian Strategic Policy Institute, November 25, 2019, p. 6.

219,102 authors who were not affiliated with our sample universities, then 25,618 who published only one article or published articles during only one year, and finally 1,389 authors who had five or more affiliation changes between 2011 and 2020 (the latter of which we find to be strongly correlated with imprecisely identified authors). The final dataset contained 290,699 distinct authors (referred to in this chapter as *researchers*) listed on 882,677 publications. Of these publications, 753,808 (85 percent) were written in English, 127,534 (14 percent) were written in Chinese, and 1,335 (1 percent) were written in other or multiple languages.

Methodology

Given that for most publication data, the most granular temporal unit of measurement associated with a publication is one year, we first condense a researcher's affiliation into *researcher-year-affiliation* observations. We then record the affiliation of a researcher's first year publishing as the researcher's *origin* and the affiliation of a researcher's most-recent year publishing as their *destination*. The terms *origin* and *destination* are applicable only within the scope of the dataset; a researcher might have published with a different affiliation before 2011 and after 2020, but these movements are not recorded. Researchers might have published a single paper with one affiliation or more in the United States and China. We refer to this instance as a *co-affiliation*. If a researcher published one paper with a U.S. affiliation and a separate paper with a PRC affiliation within a year, we refer to it as a *multiple-affiliation*. For simplicity, we refer to the presence of either type (i.e., co- or multiple-) as a *multi-affiliation*.

In our dataset, multi-affiliations can occur between US10 and PRC10 or Seven Sons universities, as well as between US10, PRC10, Seven Sons, and/or "other" universities. Because our dataset contains all publications of interest from our sample universities, we have included all instances of co-affiliations between our sample universities and universities outside our sample (i.e., "other") when they exist. With a comprehensive accounting of co-affiliations between our sample and other universities, we record these researchers as having a multi-affiliation. In contrast, our dataset does not contain all publications from all other universities in each country. Without a comprehensive list of multiple-affiliations between our sample universities and other universities, we drop all publications with "other" university affiliations when they do not contain co-affiliations with our sample universities. Without multiple-affiliations, the actual number of multi-affiliations between universities in our sample and other universities is thus likely to be greater than we present. Likewise, dropping publications that contain only other affiliations might overestimate the number of US10, PRC10, and Seven Sons affiliations that may otherwise be multi-affiliations.[13]

---

[13] Here, we provide an example of how we handle the presence of multi-affiliations between universities in our sample and other universities. A researcher who publishes a single article that lists Harvard (a US10 university) and Tianjin University (a PRC-other university) as their affiliation would have co-affiliation between a US10 and PRC-other university for the year the paper was published. If the researcher publishes one paper with Harvard and a separate paper with Tianjin, they would have a multiple-affiliation between a

There are myriad paths that a researcher can take from their origin to destination affiliations, and likewise many ways to track and categorize how a researcher's affiliation changes over time. For example, consider a researcher who publishes articles in their first year with University 1 in Country A listed as their affiliation. The following year, they publish articles with a multi-affiliation between University 1 and University 2, the latter of which is in Country B. In their third year, the researcher publishes articles with only University 1 as their affiliation. If we looked only at origin and destination (i.e., where they published in their first and final years), we would incorrectly observe that our example researcher did not move even though they published with an affiliation in Country B for one year. At the same time, if we checked only whether the researcher changed their affiliation at any point to a different country than that of their origin, we would label them only as internationally mobile and exclude the fact that they returned to their origin country in their last publishing year. What, then, is a useful way to categorize the many paths that a single researcher can take?

Our analysis places researchers into one of five affiliation-change categories, along with two summary categories.[14] We record an affiliation change when a researcher's affiliation changes between two (not necessarily consecutive) publishing years.[15] Table 3.1 provides an example affiliation path for each category over a three-year period, described in detail as follows:

1. *No Change*: The researcher does not change affiliation from any university in our sample. The exemplary case in Table 3.1 shows the researcher staying at University 1 in Country A (A1) for all three years.
2. *In-Country*: The researcher changes affiliation within their origin country at least once but is not internationally mobile within our sample. In this example, our exemplary researcher changes their affiliation to a different university in the same country in year three, from A1 to A2.

US10 and PRC-other university for that year. However, because the dataset considers only publications with a US10, PRC10, or SSons affiliation, we drop the publication with the Tianjin affiliation from the dataset, and the researcher would be listed as having only a US10 affiliation that year. If they publish one article with Brigham Young University (a U.S.-other university) and Tianjin, we would drop both publications, and the researcher would be listed as having no publication that year.

[14] The affiliation-change categories are derived, in part, from the mobility taxonomy presented in Nicolás Robinson-Garcia, Cassidy R. Sugimoto, Dakota Murray, Alfredo Yegros-Yegros, Vincent Larivière, and Rodrigo Costas, "The Many Faces of Mobility: Using Bibliometric Data to Measure the Movement of Scientists," *Journal of Informetrics*, Vol. 13, No. 1, 2019.

[15] In many cases, a researcher who changes affiliation between two universities will publish under both the previous and new universities in the year that they change affiliation. We account for this issue by checking the previous (Afft-1) and following (Afft+1) year affiliations against the "current" (Afft) year affiliations of a researcher. If Afft-1 is a single university, Afft+1 is a single university different than Afft-1, and Afft contains both the previous and following affiliation, then we consider the researcher to have changed from Afft-1 to Afft+1 in the "current" year.

TABLE 3.1

Affiliation-Change Taxonomy and Examples

Year	Non-Internationally Mobile		Internationally Mobile		
	No Change	In-Country	Migrate	Travel	Return
1	A1	A1	A1	A1	A1
2	A1	A1	B1	A1, B1	A1, B1
3	A1	A2	B1	A1, B1	A1

NOTE: A1 refers to an affiliation with University 1 in Country A and likewise for Country B and University 2.

3. *Migrate*: The researcher's destination is a single or multi-affiliation in a different country and no affiliation in the origin country. In this example, the researcher changes their affiliation to a new university in a different country in year two—from A1 to A2—and does not return to an affiliation in their origin country.

4. *Travel*: The researcher's destination is a multi-affiliation between a university in the origin country and a different country. In this example, the researcher picks up an affiliation in a different country while maintaining their former affiliation, changing from A1 to A1, B2 in year two, and maintaining both affiliations thereafter.

5. *Return*: The researcher's destination is a single affiliation in the origin country, but the researcher had a single or multi-affiliation in a different country at some point between the origin and destination. In this example, the researcher picks up an affiliation in a different country in year two—changing from A1 to A1, B2—but ultimately returns to an affiliation in their starting country in their final year—changing from A1, B2 back to A1.

Under this categorization, we do not track if a researcher has multiple international movements between the origin and destination. In this study, we were primarily focused on the number of researchers who change affiliation rather than the number of changes any researcher makes. Future research could investigate the prominence of multiple international movements.

We are, however, particularly interested in the quantitative comparisons between researchers with single and multi-affiliations, and the prominence of multi-affiliations in general. A 2019 study tracking 16 million authors from 2008 to 2015 showed that, among internationally mobile researchers, those that traveled (i.e., those whose destination was a multi-affiliation with a university in the origin and a different country) were more common than those who migrated (i.e., those whose destination was a single or multi-affiliation in a different country than the origin).[16] If a similar trend holds for our sample, it would suggest that further attention ought to be given to researchers with multi-affiliations between China and United

---

[16] Robinson-Garcia et al., 2019.

States.[17] Researchers whose origin is a multi-affiliation between the United States and China are also internationally mobile, but we address them separately from this taxonomy in a later section (see the section later in this chapter called "The Role of Multi-Affiliations").

Quantitative Assessment of Affiliation Changes

In the following sections, we begin assessing the flow of researchers by determining their origin affiliation and how frequent affiliation changes are within the sample population. We then focus on the affiliation changes of researchers whose first affiliation was in the United States or China, as well as the role of multi-affiliations. The following sections will look at the number of researchers who changed affiliation, what type of change they made, and the most-common origins and destinations between each subset of universities.

Origin Affiliation

Most researchers—192,659 (66 percent)—in the dataset had their first affiliation with a PRC10 university, followed by the Seven Sons at 49,285 (17 percent), US10 at 46,331 (16 percent), and 2,165 (1 percent) who were multi-affiliated between at least two of three university subsets or another university in the opposite country (Figure 3.1). The substantially higher number of PRC10 researchers in the dataset likely reflects that the top ten Chinese universities are larger, in terms of student enrollment and faculty, than their U.S. counterparts.

FIGURE 3.1
Count of Researcher Origin, by Group

SOURCE: Features Scopus data (Elsevier, undated-b).

---

[17] Hottenrott et al. (2021) found that from 2016 to 2019, the United States was the common host country for international researchers with multi-affiliations. At the same time, they found that more than 40 percent of U.S.-based researchers with multi-affiliations had their second affiliation with a university in China. The next most-common host for U.S.-based researchers was the United Kingdom, at roughly 10 percent.

Number of Researchers with Affiliation Changes

Within our sample population, the vast majority (89 percent) of researchers did not change their affiliation to another university within the sample subset. These researchers either maintained their first affiliation or changed their affiliation to a university outside the scope of our analysis. A total of 30,710 researchers (11 percent) changed affiliations between one and five times. The following sections will assess these researchers in further detail. Figure 3.2 depicts the count and proportion of researchers by the number of affiliation changes they made between 2011 and 2020.

Analysis of Researchers of U.S. Origin

For researchers with a U.S. origin, we computed the number of researchers who fell into our five affiliation change categories (Figure 3.3). In line with the sample population average in Figure 3.2, 41,030 U.S. researchers (89 percent) did not change affiliations within the dataset, 4,224 (9 percent) obtained only a new US10 affiliation, and 1,077 researchers (2 percent) obtained a Chinese affiliation after starting at a US10 university.

This percentage of international mobility is in line with similar research, which found 3 percent of researchers to be internationally mobile.[18] Interestingly, four out of five U.S. researchers (80 percent) who obtained a PRC affiliation migrated or traveled (i.e., their desti-

FIGURE 3.2

Authors by Number of Affiliation Changes, 2011–2020

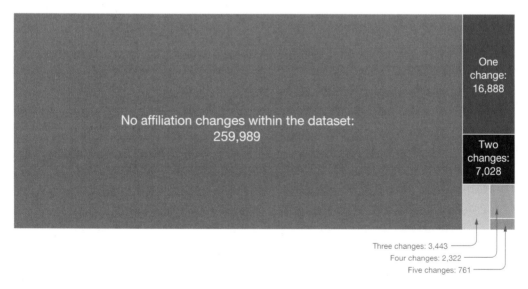

[18] Robinson-Garcia et al., 2019.

FIGURE 3.3

Affiliation Change Summary of US10-Affiliated Researchers

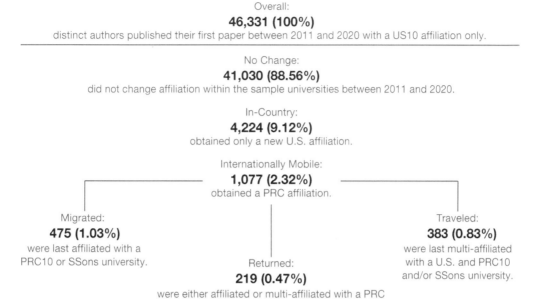

Overall:
46,331 (100%)
distinct authors published their first paper between 2011 and 2020 with a US10 affiliation only.

No Change:
41,030 (88.56%)
did not change affiliation within the sample universities between 2011 and 2020.

In-Country:
4,224 (9.12%)
obtained only a new U.S. affiliation.

Internationally Mobile:
1,077 (2.32%)
obtained a PRC affiliation.

Migrated:
475 (1.03%)
were last affiliated with a
PRC10 or SSons university.

Traveled:
383 (0.83%)
were last multi-affiliated
with a U.S. and PRC10
and/or SSons university.

Returned:
219 (0.47%)
were either affiliated or multi-affiliated with a PRC
university before returning to the US10.

SOURCE: Features Scopus data (Elsevier, undated-b).

nation had a Chinese affiliation). Only one in five (20 percent) returned to a U.S.-only affiliation by 2020 after they obtained a Chinese affiliation.

We then broke down the PRC affiliations obtained by U.S. researchers into either PRC10, Seven Sons, or PRC-other universities in the case of multi-affiliations (Figure 3.4). Perhaps unsurprisingly, most internationally mobile U.S. researchers across all categories obtained affiliations with PRC10 universities over Seven Sons universities; for every 20 U.S. researchers who obtained a PRC10 affiliation, one obtained a Seven Sons affiliation. Likewise, researchers who traveled or returned to the United States obtained multi-affiliations with PRC universities outside our sample at a rate close to that of PRC10 universities.

Looking more closely at US10 researchers whose final affiliation was with a PRC10 or Seven Sons university (and no multi-affiliation), Table 3.2 displays the most-common origins and destinations, as well as pairs between US10 and PRC10/Seven Sons universities. The most-common final affiliation, or destination, within the PRC10 was CAS, which had 99 research-

FIGURE 3.4

Breakdown of PRC Affiliations Obtained by US10 Researchers

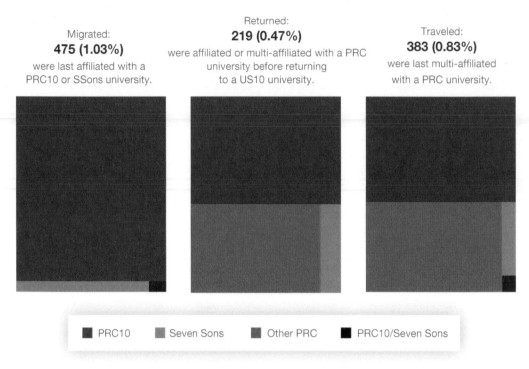

Migrated:
475 (1.03%)
were last affiliated with a
PRC10 or SSons university.

Returned:
219 (0.47%)
were affiliated or multi-affiliated with a PRC
university before returning
to a US10 university.

Traveled:
383 (0.83%)
were last multi-affiliated
with a PRC university.

■ PRC10   ■ Seven Sons   ■ Other PRC   ■ PRC10/Seven Sons

SOURCE: Features Scopus data (Elsevier, undated-b).

TABLE 3.2

Top Nodes and Pairs of US10 Researcher Affiliation Changes

Affiliation	Top Destinations	Top Pairs	Top Origins
PRC10	1. CAS: 99 2. Tsinghua: 58 3. Peking: 54	1. MIT to CAS: 20 2. California Institute of Technology to CAS: 13 3. Harvard to CAS: 12	1. MIT: 86 2. UC Berkeley: 63 3. Stanford: 58
Seven Sons	1. HIT: 12 2. BIT: 6 3. Beihang: 5	1. MIT to BIT: 6 2. Rest tied: 1	1. MIT: 9 2. Stanford: 4 3. UC Berkeley: 3

SOURCE: Features information from Elsevier, undated-b.
NOTE: Author counts might include multi-affiliations counted elsewhere in table.

ers who were first affiliated with a US10 university.[19] For the Seven Sons, the most-common destination was HIT, which had 12 former US10-affiliated researchers. The most-common origin of those going toward the PRC10 and Seven Sons was MIT. These most-common destinations and origins also made up some of the most-common pairs of affiliation changes between the US10, the PRC10, and the Seven Sons.

Analysis of Initially PRC-Affiliated Researchers

We now turn to the flow from PRC10 and Seven Sons to U.S. universities. Like the previous section, we computed the number of researchers who fell into our five affiliation change categories. Almost 90 percent of Chinese researchers did not change affiliations within the dataset: a rate nearly identical to that of U.S. researchers. Of the remainder, 17,594 (7 percent) obtained only a new PRC affiliation, while 6,876 researchers (3 percent) obtained a U.S. affiliation.

While 80 percent of US10 researchers who obtained a PRC affiliation migrated or traveled (i.e., had a single or multi-affiliation with a PRC university as their destination), just over 50 percent of internationally mobile PRC10 and Seven Sons researchers migrated or traveled. The other half of PRC10 and Seven Sons researchers returned to a PRC affiliation, whereas only 20 percent of US10 researchers returned to a U.S. affiliation. This disparity might be the result of the number of Chinese graduate students who first publish while studying in the United States before returning to China after graduation.

To highlight the differences between PRC10 and Seven Sons researchers who obtained a U.S. affiliation, Figure 3.5 breaks down the three categories of international mobility like that of Figure 3.3. Of those whose final affiliation was a US10 university, 705 researchers (94 percent) had their first affiliation with a PRC10 university, while only 6 percent came from the Seven Sons. Given that the PRC10 and Seven Sons researchers made up 79 percent and 20 percent of the PRC population in the data, respectively, the PRC10 swung above its proportional weight in researchers adopting a US10 affiliation.

PRC10 and Seven Sons researchers who either traveled or returned to the PRC after having a multi-affiliation with a U.S. university had a different affiliation pattern than their US10 counterparts who we discussed previously. Whereas the US10 researchers who traveled or returned after having a multi-affiliation with a PRC university primarily favored affiliations with PRC10 universities, both PRC10 and Seven Sons researchers overwhelmingly had multi-affiliations with U.S.-other (non-US10) universities. For instance, of Seven Sons researchers who obtained a U.S. affiliation before returning to a Chinese affiliation, 532 had multi-affiliations with U.S.-other universities while 50 had affiliations or multi-affiliations with US10 universities. A similar, but slightly higher, ratio holds true for PRC10 researchers as well, as can be seen in Figure 3.6. As a reminder, the count of multi-affiliations with "other"

---

[19] We acknowledge that CAS is an agglomeration of many research institutes and not necessarily the equivalent of other universities listed here. For example, our sample has 54,834 researchers with CAS as their origin, the highest volume of any university we considered. The U.S. university with the highest volume, MIT, had 8,654 researchers.

FIGURE 3.5

Affiliation Change Summary of PRC10 and Seven Sons–Affiliated Researchers

Overall:
242,395 (100%)
distinct authors published their first paper between 2011 and 2020 with a PRC10
or Seven Sons affiliation only.

No Change:
217,925 (89.90%)
did not change affiliation within the sample universities between 2011 and 2020.

In-Country:
17,594 (7.26%)
obtained only a new PRC10 and/or Seven Sons affiliation.

Internationally Mobile:
6,876 (2.84%)
obtained a U.S. affiliation.

Migrated:
753 (0.31%)
were last affiliated with a
US10 university.

Returned:
3,239 (1.34%)
were either affiliated or multi-affiliated with a U.S. university
before returning to a PRC10 or Seven Sons affiliation only.

Traveled:
2,884 (1.19%)
were last multi-affiliated
between a U.S. and PRC
university.

SOURCE: Features Scopus data (Elsevier, undated-b).

universities here includes only co-affiliations because of the limitations of our data. In turn, the counts of "other" multi-affiliations are likely to be greater than the computations we present. This note further suggests that significant affiliation flows from the PRC10 and Seven Sons to the United States occur outside the scope of the current dataset.

Turning to the university level, Table 3.3 displays the most-common origins and destinations from PRC10 and Seven Sons to US10 universities, as well as the most-common pairs. The most-common destination, or final affiliation, for PRC10 researchers within the US10 was MIT, with 139 researchers. For the Seven Sons, the most-common destination was Stanford University, which had ten former Seven Sons–affiliated researchers. The most-common origin was Tsinghua University for PRC10 researchers and HIT for the Seven Sons. The most-common origins and destinations also made up the most-common pairs of affiliation changes between the US10 and the PRC10 and Seven Sons. Of note, the top US10 destinations for PRC10/Seven Sons researchers—MIT, Stanford University, UC Berkeley—are the most-common origins for US10 researchers who had PRC10 and Seven Sons destinations, as shown in Table 3.2. The top PRC10 and Seven Sons origin universities here are the same as the top destinations for internationally mobile US10-researchers in our sample. The similarity between the nodes from the US10

FIGURE 3.6

Breakdown of U.S. Affiliations Obtained by PRC10 and Seven Sons Researchers

SOURCE: Features Scopus data (Elsevier, undated-b).

TABLE 3.3

Top Origins, Destinations, and Origin-Destination Pairs of Internationally Mobile PRC10 and Seven Sons Researchers with US10 Universities

Affiliation	Top Destinations	Top Pairs	Top Origins
PRC10	1. MIT: 147 2. Stanford: 126 3. UC Berkeley: 109	1. Tsinghua to MIT: 49 2. Tsinghua to UC Berkeley: 36 3. Peking to Stanford: 36	1. Tsinghua: 188 2. Peking: 178 3. CAS: 89
Seven Sons	1. Stanford: 10 2. UC Berkeley: 8 3. MIT: 8	1. HIT to MIT: 5 2. HIT to Stanford: 4 3. BIT to Stanford: 4	1. HIT: 22 2. Beihang: 10 3. BIT: 9

SOURCE: Features Scopus data (Elsevier, undated-b).
NOTE: Author counts might include multi-affiliations counted elsewhere in table.

to PRC10 and Seven Sons and vice versa suggests there might be well-established pathways or networks between these U.S. and Chinese universities.[20]

The Role of Multi-Affiliations

As discussed earlier, we sought to explore the prominence of multi-affiliations among internationally mobile researchers. First, internationally mobile researchers obtained multi-affiliations more often than they directly migrated (i.e., obtaining only a single affiliation). Of the 685 U.S. and 6,876 PRC researchers who obtained an affiliation in the opposite country, 86 percent either obtained or finished with a multi-affiliation while 14 percent changed and never obtained a multi-affiliation. Moving from one university in one country directly to another university in another country was less common than obtaining a multi-affiliation between universities of both countries at some point.

We also explored whether researchers who obtained a multi-affiliation at some point were more likely to migrate than those who directly migrated—that is, whether obtaining a multi-affiliation was a common intermediate step to migration. Among single-origin researchers who migrated, however, only 17 percent had obtained a multi-affiliation prior to their destination, indicating that direct migration was more common than obtaining a multi-affiliation before migrating and that multi-affiliations were not a common intermediate step to migration. Instead, most researchers who obtained multi-affiliations at some point ultimately traveled or returned rather than migrated. Only 3 percent of the 86 percent who obtained or finished with a multi-affiliation had their destination in the opposite country, while 49 percent returned to their origin country and 48 percent remained with multi-affiliation. Figure 3.7 depicts the affiliation pathway of internationally mobile researchers from their origin country to their destination by proportion, with the middle node representing those researchers who obtained a multi-affiliation prior to their destination.

There are also country-specific dynamics at play. Of the 6,876 internationally mobile researchers from the PRC10 and Seven Sons, 6,170 (90 percent) obtained a multi-affiliation. In contrast, 685 (64 percent) of 1,077 internationally mobile researchers from the US10 obtained a multi-affiliation with a PRC university. That is, PRC10 and Seven Sons researchers were

---

[20] Indeed, there is a dense social mesh between American and Chinese students and scholars. Since the 2009–2010 academic year, Chinese nationals have been the largest group of international students studying in the United States, with 290,086 Chinese students as of 2021–2022. At the same time, the Chinese government has opened more than 100 Confucius Institutes across U.S. schools since 2004 and sponsors numerous exchange and talent programs, Similarly, 133 U.S. universities operated 225 cooperative education programs with Chinese universities as of 2016. The United States and China are also one another's most significant bilateral partner for academic collaboration, although collaborations are decreasing (as will be discussed in Chapter 4). For more on U.S.-Chinese academic ties, see Anastasya Lloyd-Damnjanovic and Alexander Bowe, *Overseas Chinese Students and Scholars in China's Drive for Innovation*, U.S. China Economic and Security Review Commission, October 7, 2020; U.S. Government Accountability Office, *U.S. Universities in China Emphasize Academic Freedom but Face Internet Censorship and Other Challenges*, U.S. Government Printing Office, GAO-11-502, June 2011; Institute of International Education, "All Places of Origin," spreadsheet, 2022.

FIGURE 3.7

Affiliation Change Pathways for Single-Origin, Internationally Mobile Researchers

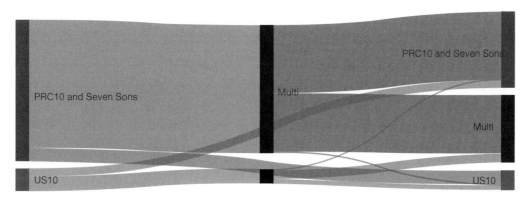

SOURCE: Features Scopus data (Elsevier, undated-b).
NOTE: "Multi" refers to "Multi-affiliation."

70 percent more likely than their US10 counterparts to obtain multi-affiliations. There is also a difference between countries in the destinations of researchers who obtained a multi-affiliation. Of the 6,170 PRC10 and Seven Sons researchers who had a multi-affiliation, 119 (2 percent) ultimately migrated to a US10 affiliation, 3,167 (51 percent) returned to a PRC10 or Seven Sons affiliation, and 2,884 (47 percent) traveled. Of the 685 US10 researchers who had a multi-affiliation, 98 (14 percent) migrated to a PRC10 or Seven Sons affiliation, 204 (30 percent) returned to a US10 affiliation, and 383 (56 percent) traveled. The relative proportions of each group are presented in Figure 3.8.

Several differences stand out. The higher proportion of US10 researchers who remained travelers or migrated instead of returning to a US10 affiliation shows that U.S. researchers who obtained a multi-affiliation were twice as likely as their PRC10 and Seven Sons counterparts to maintain their PRC affiliation than to relinquish it. In contrast, PRC10 and Seven Sons researchers were just as likely to return to a Chinese affiliation as they were to maintain their multi-affiliation or migrate to a US10 university. This disparity might be the result of Chinese graduate students who first published while in U.S. doctoral programs then migrated back to China or maintained affiliations across the Pacific after completing their doctoral studies. The relatively higher proportion of US10 researchers who migrated following a multi-affiliation might speak to the same phenomenon.

Much of the assessment in this chapter thus far has focused on researchers who had single start and end affiliations. However, those who had a multi-affiliation as their origin made a small but significant contribution to the flow of author affiliation changes. Figure 3.9 shows a breakdown by affiliation of the 2,156 researchers whose origins were multi-affiliations. Excepting multi-affiliations between the PRC10 and Seven Sons, which were counted under the PRC figures above, the most-common starting multi-affiliation pairs were between the

FIGURE 3.8

Destination of Single-Origin, Internationally Mobile Researchers Who Obtained a Multi-Affiliation, by Proportion

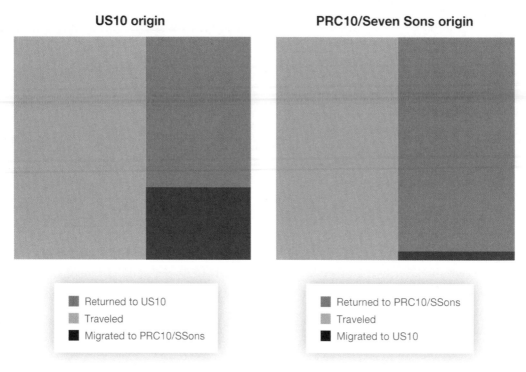

SOURCE: Features Scopus data (Elsevier, undated-b).

PRC10 and U.S.-other universities, US10 and PRC-other universities, and Seven Sons and U.S.-other universities. On one hand, the frequency of "other" multi-affiliations again suggests a significant amount of affiliation change from inside the sample group to outside our sample. On the other, it shows how relatively uncommon multi-affiliations between the university subsets in our sample are.

Finally, we turn to multi-affiliations at the university level across all researchers in our sample. Table 3.4 highlights the top five multi-affiliation pairs between the US10 and PRC10, as well as between the US10 and Seven Sons. The most prominent multi-affiliate from the US10 was MIT, leading both lists with Tsinghua and HIT, respectively. MIT also had the most multi-affiliations with PRC10 universities, with 325 researchers holding a MIT-PRC10 multi-affiliation at some point in the dataset. Conversely, UCB had the highest total number of Seven Sons multi-affiliations at 46 researchers. Among the PRC10, Tsinghua and CAS led in overall multi-affiliations with 368 and 355, respectively, while HIT had the most multi-affiliations with US10 among Seven Sons universities at 80 researchers.

FIGURE 3.9

Count of Authors with a Multi-Affiliation Origin

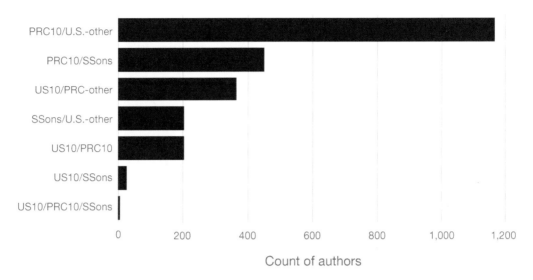

SOURCE: Features Scopus data (Elsevier, undated-b).

TABLE 3.4

Top Multi-Affiliation Pairs by Count of Researchers

US10 – PRC10		US10 – Seven Sons	
Pair	**Authors**	**Pair**	**Authors**
Tsinghua MIT	89	HIT MIT	18
Tsinghua UC Berkeley	69	HIT Yale	14
Nanjing Yale	67	HIT Columbia	12
CAS MIT	63	Harbin UC Berkeley	11
Tsinghua Stanford	53	HIT/BIT UC Berkeley	1

SOURCE: Features Scopus data (Elsevier, undated-b).

A Qualitative Assessment: Impact and Knowledge

In the above analysis, the only qualitative information known for researchers was that they had published in at least one of Scopus's Physical Sciences subject areas and that they were affiliated with a university in the US10, PRC10, or Seven Sons. Although filtering the data on these criteria focused our analysis on researchers more likely to be involved with China's military-civil fusion or to have a greater research influence more generally, there are ways to further assess the qualitative aspects of the researcher groups in question. In fact, a more nuanced look at knowledge flows and research impact can show both benefits and costs to researcher-affiliation changes.[21]

In the sections below, we first look at research impact by means of average citations per author. Although an incomplete measure of research influence, citations show how widely a researcher's knowledge is read and used as a basis for further research. We then turn to Scopus-indexed keywords used by each author to assess what areas of knowledge accompany an internationally mobile researcher. Scopus generates a standardized list of keywords for each publication entered into its database, allowing for a systematic accounting of topics covered by a researcher across publications. By attaching this keyword accounting to a researcher's affiliation status, we can see, by proxy of the keyword, what bodies of knowledge move with internationally mobile researchers.

Research Influence: Citations

To assess the research influence of internationally mobile researchers within our sample, we computed the average number of citations per researcher within our affiliation-change taxonomy (Figure 3.10). Although US10, PRC10, and Seven Sons researchers shared the same pattern of citations per author across affiliation-change groups, US10 researchers had greater citations per author than their PRC10 and Seven Sons counterparts in every category. For both U.S. and PRC researchers, internationally mobile researchers had a greater number of citations on average than nonmobile researchers in our sample (Figure 3.10). We also found that returnees (i.e., researchers who returned to their initial country after publishing in the opposite country) had roughly three times the number of citations on average than that of the next leading internationally mobile category. Internationally mobile researchers within our sample therefore had greater research influence (by proxy of citations) than nonmobile researchers, while returnees had the most significant impact within the internationally mobile.[22]

---

[21] While one school of thought views international mobility in the zero-sum terms of *brain drain* and *brain gain* (i.e., a country either loses or gains researchers), another sees international mobility as a mechanism by which researchers can enter international collaboration networks and thereby increase advancement and awareness of scientific research across countries. Neither is entirely correct in all cases of international mobility.

[22] We included all publications for researchers in our calculation. We did not distinguish publications before or after a researcher changed their affiliation but rather included all publications for each researcher

FIGURE 3.10
Citation per Author by Affiliation-Change Categories

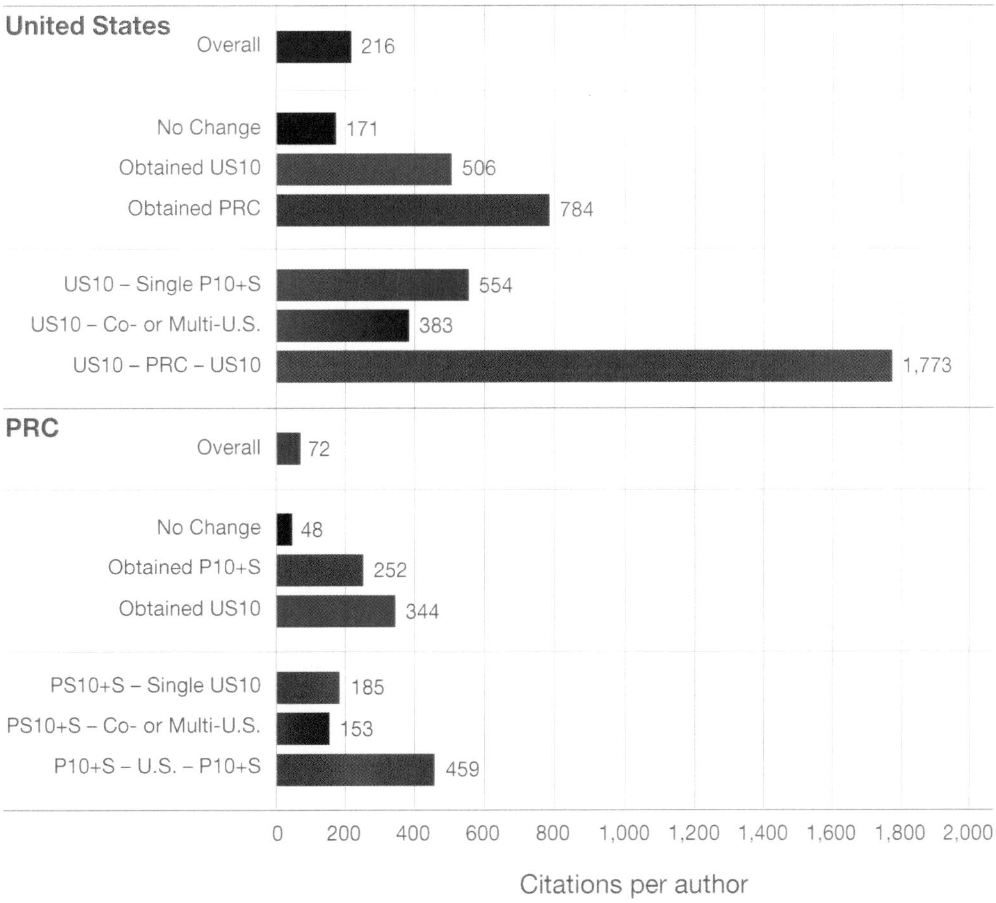
United States
Overall
216
No Change
171
Obtained US10
506
Obtained PRC
784
US10 – Single P10+S
554
US10 – Co- or Multi-U.S.
383
US10 – PRC – US10
1,773
PRC
Overall
72
No Change
48
Obtained P10+S
252
Obtained US10
344
PS10+S – Single US10
185
PS10+S – Co- or Multi-U.S.
153
P10+S – U.S. – P10+S
459
0
200
400
600
800
1,000
1,200
1,400
1,600
1,800
2,000
Citations per author

SOURCE: Features Scopus data (Elsevier, undated-b).
NOTE: "P10+S" is a shortened version of "PRC10 and Seven Sons."

Areas of Knowledge: Keywords

To assess the most-common areas of knowledge accompanying internationally mobile researchers between the subset university groups of our sample, we assessed the prominence of index keywords associated with these internationally mobile researchers. Here, keywords provide a sense of what scientific topics a researcher studies, shares, and learns more about with collaborators during or after their mobility. On one hand, this assessment can highlight

---

in each affiliation change category. Further research could evaluate the number and citations of researchers before, during, and after specific affiliation changes to better understand the effect of the affiliation changes.

researchers involved in "civil and/or military fusion" technologies. On the other hand, it can show the areas of greatest international knowledge flows that both countries can leverage and utilize. In either case, tracking the keywords paired with different groups of internationally mobile researchers provides a tool to monitor specific knowledge flows. Depending on the user, this information could point to flows that policymakers might want to encourage, dampen, or otherwise leverage in their decisionmaking.

We first found the five most-frequently used index keywords (i.e., the frequency of keywords across all a researcher's publications, summed for all researchers) of researchers who obtained an affiliation in a different group. We then calculated the top five keywords associated with the most researchers (i.e., the number of researchers who use a given keyword) who obtained an affiliation in a different group. The former is primarily influenced by the number of publications with a keyword, allowing a handful of prolific researchers publishing on the same topic to give more weight to a given keyword. The latter favors the number of researchers using a keyword irrespective of how much they publish. Affiliation groups are simplified for this analysis; any researcher who obtained an affiliation in a different group than their origin, regardless of destination or multi-affiliation, is counted as a single affiliation category. The results are found in Table 3.5.

Interestingly, there are common keywords across almost all affiliation categories. *Graphene*—a nanocarbon first properly isolated and characterized in 2004 with potential application in semiconductors, batteries, and quantum computers—appears in all US10-to-PRC10, PRC10-to-US10, and Seven Sons–to-US10 lists. *Scanning electron microscopy*—a tool used broadly in materials science, as well as in the production of semiconductors and microchips—also appears in the same groups. *Lithium* appears in all affiliation categories but US10-to-PRC10, whereas *tellurium compounds*—a material used in copper and steel alloys, along with solar panels and some semiconductors—is present only in the US10-to-PRC10 category.

Broadly, the US10-to-PRC10 category and its converse share many common elements: *graphene, metabolism, electrodes, density functional theory*, and others. However, the US10-to–Seven Sons category and its converse depart more significantly from one another. Out of the ten keywords listed for each category, only *graphene, optimization*, and *lithium* appear in both categories. US10-to–Seven Sons keywords include *magnetism, geometry, polymers*, and *aluminum*; Seven Sons–to-US10 keywords include *electrodes, scanning electron microscopy*, and *numerical methods*.

The structure of our keyword analysis also allows us to compute the counts for specific keywords. Table 3.6 represents our calculation of the frequency of and count of researchers using the keyword *radar*. We see a handful of researchers in each affiliation category had several publications with *radar* as an indexed keyword with the exception of US10-to–Seven Sons, which had no researchers published on *radar*. A similar table could be produced for any keyword of interest and could include dual-use and military-relevant terms or areas of civilian interest.

TABLE 3.5

Top Keywords Associated with Internationally Mobile Researchers

Affiliation	Keywords (Frequency)	Keywords (Researchers)[a]
US10 + PRC10	Graphene (667) Metabolism (503) Electrodes (428) Density functional theory (412) Tellurium compounds (405)	Graphene (152) Metabolism (151) Surface property (150) Chemical structure (137) Scanning electron microscopy (132)
US10 + Seven Sons	Optimization (25) Magnetism (22) Magnetoplasma (22) Decision making (21) Lithium (21)	Geometry (9) Electrons (8) Ions (8) Polymers (8) Aluminum compounds (7)
PRC10 + US10	Graphene (1,185) Electrodes (961) Lithium (812) Metabolism (759) Scanning electron microscopy (692)	Scanning electron microscopy (317) Graphene (311) Metabolism (296) Chemical structure (288) Density functional theory (258)
Seven Sons + US10	Electrodes (164) Graphene (157) Lithium-ion batteries (85) Oxidation (84) Scanning electron microscopy (83)	Numerical methods (33) Scanning electron microscopy (32) Computer simulation (30) Graphene (30) Optimization (30)

SOURCE: Features data from Elsevier, undated-b.
NOTE: Table excludes the following broadly descriptive keywords: *article, animal, carbon, chemistry, China, controlled study, human, oxygen, priority journal, procedures, temperature, unclassified drug, United States.*

[a] Researchers with multi-affiliations might be included across affiliation categories.

TABLE 3.6

Use of *Radar* as a Specific Keyword Search

Affiliation	Keywords (Frequency)	Keywords (Authors)
US10-to-PRC10	19	7
US10-to-Seven Sons	0	0
PRC10-to-US10	22	11
Seven Sons-to-US10	12	4

SOURCE: Features data from Elsevier, undated-b.
NOTE: Researchers with multi-affiliations might be included across affiliation categories.

Conclusion

Our analysis of affiliation changes between the US10, PRC10, and Seven Sons universities showed that fewer than 3 percent of both U.S. and Chinese researchers were internationally mobile between the universities in our sample. For researchers with a US10 origin who obtained a PRC affiliation, we found mobility toward PRC10 universities was 20 times that of Seven Sons universities. We also found that US10 researchers migrated to a PRC affiliation or held a multi-affiliation more often than they returned to the United States. In contrast, PRC10 and Seven Sons researchers returned to China or maintained a multi-affiliation more often than they migrated to a U.S. affiliation. For the PRC10 and Seven Sons researchers who obtained a U.S. affiliation, a greater number obtained multi-affiliations with universities outside the US10 than with a US10 institution, indicating that there are substantial flows with universities outside our sample. Among individual universities, the most-common origins and destinations of internationally mobile researchers in both directions were MIT, Stanford University, and UC Berkeley for the US10; CAS, Peking University, and Tsinghua University for the PRC10; and Beihang University, BIT, and HIT for the Seven Sons.

Of the 685 U.S.-based and 6,876 PRC-based internationally mobile researchers with a single country origin in our sample, 86 percent either obtained or finished with a multi-affiliation. Internationally mobile researchers with a single origin in our sample were more likely to obtain multi-affiliations with a university in the opposite country (at some point) than to directly migrate. At the same time, researchers who obtained a multi-affiliation between countries were more likely to maintain their multi-affiliation or return to their origin country than they were to migrate. The most-common multi-affiliation university pairs between the US10 and PRC10 and the US10 and Seven Sons featured the most-common origin and destination universities listed previously, along with Yale University, Columbia University, Nanjing University, and Harbin Engineering University.

Our qualitative assessment showed that internationally mobile researchers had greater researcher impact on average than those who did not obtain international affiliations by proxy of citations per author. Among internationally mobile researchers, returnees had the greatest impact. Tracking the keywords of internationally mobile researchers revealed that certain areas of research moved ubiquitously between groups, particularly in the areas of semiconductors, batteries, and microchip production. We also demonstrated that our keyword analysis could be used to assess international mobility in user-specified topics of interest.

Our methodology in the chapter assessed a sample of researchers from the United States and China over a ten-year period. However, it could readily be expanded to a larger scope. With a more comprehensive dataset of U.S. and PRC researchers, it could provide a more detailed, university-level analysis of internationally mobile researcher flows. The methodology could also expand to investigate year-on-year affiliation changes, which could discern trends over time across all levels of analysis presented above. Further research could uncover more-qualitative factors about international mobility, such as type of affiliation (e.g., short-term appointment or long-term move), via the duration between affiliation changes and assessment of funding sources where available. Overall, this assessment has shown that publication data can track the affiliation changes of researchers between the United States and China at the university level and assess qualitative details of internationally mobile researchers, such as influence and knowledge.

CHAPTER 4

Exploration of the Potential Risks and Benefits of U.S.-China Collaboration in Aerospace Research

In the field of aerospace engineering, research collaboration between the United States and China has increased in recent years (see Figure 4.1).[1] Without further analysis, this trend says little about the potential risks or benefits of this collaboration to U.S. national interests.[2] On one hand, increased collaboration with China might represent a means of academic collection: the intentional and state-directed collection of scientific knowledge and know-how from a country's academic institutions by a foreign actor. The PRC's academic-collection efforts are well documented and constitute a potentially important source of loss of U.S.-produced scientific knowledge.[3] Even when not prone to academic collection, scientific collaboration on sensitive topics constitutes a risk if U.S.-generated knowledge is used to advance China's military modernization objectives. On the other hand, international collaboration is linked to greater scientific impact,[4] and, historically, scientific diplomacy has been a productive means of improving international relations.[5] Hottes et al. point out

---

[1]   Evidence presented by Wagner and Cai indicates that there was a (countertrend) decrease in overall U.S.-China scientific collaboration in 2021 (Caroline S. Wagner and Xiaojing Cai, "Drop in China-USA International Collaboration," *ISSI Newsletter*, Vol. 18, No. 1, March 2022). This may be because of recent political tension between the countries or increased scrutiny by the U.S. government of Chinese researchers working in the United States.

[2]   Neither does the observation of this trend say anything about the determinants of collaboration growth. Possible determinants include China's overall increasing sophistication in aerospace engineering, an increase in the number of Chinese nationals attending college and graduate school in the United States, and government funding of international collaborative research.

[3]   Alex Joske, *Picking Flowers, Making Honey*, Australian Strategic Policy Institute, 2018; Portman and Carper, 2019.

[4]   Vicente P. Guerrero Bote, Carlos Olmeda-Gómez, and Félix de Moya-Anegón, "Quantifying the Benefits of International Scientific Collaboration," *Journal of the American Society for Information Science and Technology*, Vol. 64, No. 2, 2013.

[5]   Guerrrero Bote et al., "Quantifying the Benefits of International Scientific Collaboration," *Journal of the American Society for Information Science and Technology*, Vol. 64, No. 2, 2013; Vaughan Turekian, "The Evolution of Science Diplomacy," *Global Policy*, Vol. 9, No. 53, 2018, pp. 5–7.

FIGURE 4.1

U.S.-China Collaborations on Aerospace Engineering Research, 2001–2020

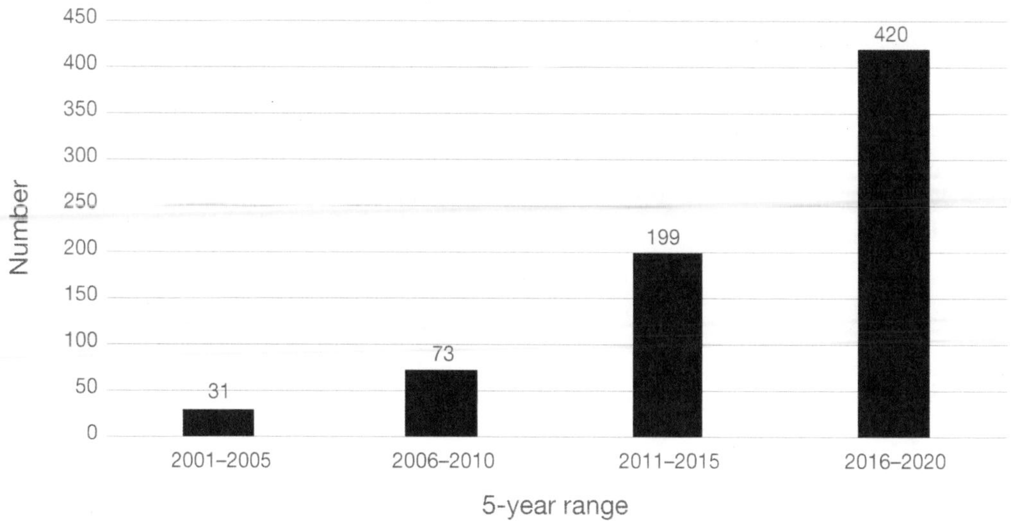

SOURCE: Features WOS data (Clarivate, undated-b).

that international scientific collaboration also constitutes a channel by which the United States might promulgate research norms, values, and ethics abroad.[6] The objective of this chapter is to consider in greater detail the data underlying the trend of increased collaboration in aerospace engineering with an eye toward understanding potential sources of risk and benefit to the United States.[7]

Data

The scientific publication data used here come from the Web of Science (WOS), a database of more than 90 million records and containing the top journals and conference proceedings from all major scientific and engineering fields.[8] To build an aerospace engineering

---

[6] Alison K. Hottes, Marjory S. Blumenthal, Jared Mondschein, Matthew Sargent, and Caroline Wesson, *International Basic Research Collaboration at the U.S. Department of Defense: An Overview,* RAND Corporation, RR-A1579-1, 2023.

[7] The Department of Defense is aware of the potential benefit of international scientific engagement. The department's 2020 engagement strategy describes methods to access and leverage capabilities and knowledge developed abroad (U.S. Department of Defense, *DOD International Science and Technology Engagement Strategy,* 2020). Furthermore, individual Department of Defense offices, such as the Office of Naval Research Global and the Air Force Office of Scientific Research's International Office, focus on international research coordination and collaboration.

[8] Clarivate, undated-b.

publication dataset, we filter the WOS data using the "Engineering, Aerospace" WOS Category grouping for the 2001–2020 period. In the analysis to follow, an *international collaboration* is defined as a publication with two or more authors who are affiliated with organizations based in different countries.

We consider two subsets of Chinese universities with ties to the PLA.[9] First, we consider the union of two groups: the Seven Sons of National Defense and the Seven Sons of Arms Industry.[10] These universities, many of which were once formally subordinate to the State Administration for Science, Technology and Industry for National Defense, have particularly strong ties to the PLA. While these Sevens Sons universities also conduct substantial civilian research, they consider themselves to be "defense science, technology and industry work units" and instruments of the "defense system."[11] Second, we consider PLA-affiliated research organizations: a set of universities and research organizations that are directly operated by the PLA.[12] These organizations sit within the PLA organizational hierarchy and directly serve its modernization needs. While collaboration by a U.S.-based author with a Seven Sons– or PLA-affiliated university does not, in itself, constitute a harm to national security, we posit here that because these sets of Chinese organizations have close and formal ties to the PLA, such collaborations can reasonably be assumed to constitute a potential risk that is worth monitoring.

---

[9] We rely on the lists of universities provided in the following two sources: James Mulvenon and Chenny Zhang, "Targeting Defense Technologies," in William C. Hannas and Didi Kirsten Tatlow, eds., *China's Quest for Foreign Technology: Beyond Espionage,* Routledge, 2020, pp. 92–110; and Joske, 2019.

[10] The universities regarded as the Seven Sons of National Defense are BIT, Beijing University of Astronautics and Aeronautics, Harbin Engineering University, HIT, Nanjing University of Astronautics and Aeronautics, Nanjing University of Science and Technology, and Northwest Polytechnical University. The universities regarded as the Seven Sons of Arms Industry are BIT, Changchun University of Science and Technology, Chongqing University of Technology, Nanjing University of Science and Technology, North University of China, Shenyang Ligong University, and Xi'an Technological University.

[11] Joske, 2019.

[12] This set of universities are those designated as "military" by the China Defense University tracker. At the time of analysis (March 2022), this group consisted of the following universities: National University of Defense Technology, National Key Laboratory for Parallel and Distributed Processing, PLA University of Science and Technology, PLA Information Engineering University, Zhengzhou Information Science and Technology Institute, Zhengzhou Institute of Surveying and Mapping, Air Force Engineering University, Second Artillery Engineering College, Xi'an Research Institute of High Technology, Academy of Armored Force Engineering, Academy of Equipment Command and Technology, National Digital Switching System Engineering and Technological Research Center, Northwest Institute of Nuclear Technology, China Aerodynamics Research and Development Center, Naval University of Engineering, and PLA Electronic Engineering Institute. See Australian Strategic Policy Institute International Cyber Policy Centre, "China Defence Universities Tracker: Beihang University," webpage, undated.

Potential Risks Associated with Scientific Collaboration with China

Collaboration with the Seven Sons Universities

We find that U.S. collaboration with the Seven Sons universities in aerospace publishing is common; in the most recent period, half of all U.S. collaborations (i.e., 210 of 420 collaborations) with China were with this set of universities (see Figure 4.2). The Chinese organization with the highest number of coauthored publications with U.S.-based organizations was Beijing University of Aeronautics and Astronautics, one of the Seven Sons of National Defense and classified by the Australian Strategic Policy Institute as *very high risk* based on "its top-secret security credentials, high number of defence laboratories and defence research areas, and strong relationship with the defence industry."[13]

Figure 4.3 depicts the collaboration network of U.S. and Seven Sons universities for the 2011–2015 period and the 2016–2020 period. Blue nodes are U.S.-based organizations and red nodes are Chinese. Comparing the two nodes shows that in the more-recent period, more U.S. organizations collaborated with Seven Sons universities. The network graphs in Figure 4.3 show that novel U.S-China links were created during the 2016–2020 period, and there were 47 U.S.-Seven Sons collaborative dyads during the 2011–2015 period and 86 during the 2016–2020 period.

FIGURE 4.2

U.S. Collaboration with the "Seven Sons" Universities, 2001–2020

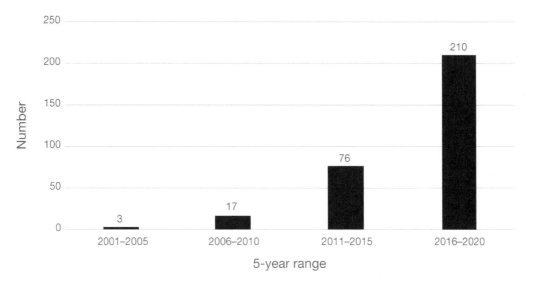

SOURCE: Features WOS data (Clarivate, undated-b).

---

[13] Australian Strategic Policy Institute International Cyber Policy Centre, 2021.

FIGURE 4.3

Collaboration Network, 2011–2020

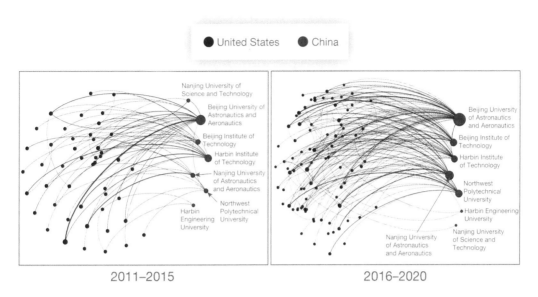

SOURCE: Features WOS data (Clarivate, undated-b).
NOTE: To increase the clarity of the visualizations, the blue nodes are not labeled.

Collaboration with PLA-Affiliated Organizations

We found that collaboration with PLA-affiliated organizations on aerospace research is relatively uncommon (see Figure 4.4). However, any U.S.-PLA collaboration on the topic of aerospace engineering might be of interest to national security policymakers and scholars.

Although collaboration with PLA-affiliated organizations on aerospace research is relatively uncommon, it has increased over time. Figure 4.5 depicts the collaboration network of U.S. and PLA-affiliated organizations for the 2011–2015 period and the 2016–2020 period.

FIGURE 4.4

U.S. Collaboration on Aerospace Research with PLA-Affiliated Organizations, 2001–2020

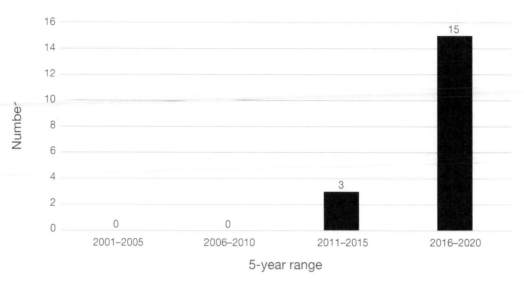

SOURCE: Features WOS data (Clarivate, undated-b).

FIGURE 4.5

Aerospace-Research Collaboration Network, 2011–2020

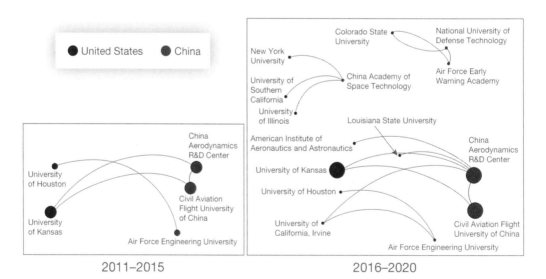

SOURCE: Features WOS data (Clarivate, undated-b).

Analysis of the Content of U.S.-Chinese Aerospace Collaborations

U.S. organizations are collaborating with Chinese organizations with ties to the PLA on dual use and even overtly military research. Table 4.1 depicts the number of aerospace engineering publications coauthored by U.S. and Chinese organizations for a set of dual use and military-relevant terms. To populate the table, we searched the titles and abstracts of the publications for a given term (a practice known as *target tracking*). The counts in the table represent the number of articles published via U.S.-Chinese collaboration that contain the term within the title or abstract. The data represented in the table indicate that there is a nontrivial amount of collaboration on hypersonic systems (e.g., the terms *hypersonic* and *scramjet*) between U.S. organizations and Chinese organizations with ties to the PLA, as well as collaboration on other topics of military relevance.

TABLE 4.1

Number of U.S.-Chinese Aerospace Research Collaborations by Category and Topic, 2001–2020

Term	China (All Organizations)	Seven Sons of National Defense and Arms Industry	PLA-Affiliated Research Organizations
Radar	73	16	2
Satellite	60	15	1
Autonomous	41	19	0
Target tracking	26	5	0
Hypersonic	24	16	2
Target detection	23	11	0
Missile	18	13	1
Scramjet	10	7	4

SOURCE: Features WOS data (Clarivate, undated-b).

The Benefits of Collaboration with China

We measure the benefits of collaboration with China in two ways: research impact and inter-disciplinarity.[14] Our proxy for research impact is citations. Our proxy for interdisciplinarity is the number of WOS topic categories that are assigned to a given publication. We find that publications produced via U.S.-China collaboration have, on average, higher impact and are more interdisciplinary.

Research Impact

Citations are a common proxy for publication impact. Highly cited publications are considered high-impact because they have been frequently used by other researchers as they document their own scientific advancements. During each of the last five years, aerospace publications written by teams composed of researchers from the United States and China have been cited more frequently than the average aerospace publication (see Figure 4.6).[15]

FIGURE 4.6

Citation Percentile, All Aerospace Publications and U.S.-China Collaborations, 2016–2020

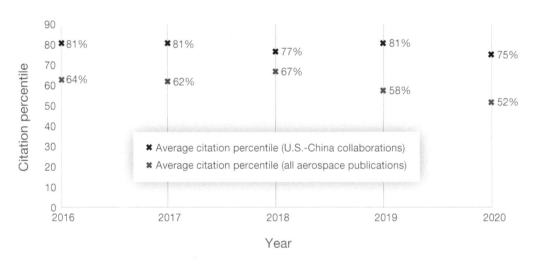

SOURCE: Features WOS data (Clarivate, undated-b).

---

[14] The case for the scientific or social utility of interdisciplinarity owes to the hypothesis that the integration of knowledge, tools, data, concepts, and perspectives from often stovepiped scientific disciplines is essential to solving the thorniest scientific and social problems. For a strong example of this argument, see Heidi Ledford, "How to Solve the World's Biggest Problems," *Nature*, Vol. 525, No. 7569, 2015.

[15] All citation percentile differentials are statistically significant at the 0.05 level. We use citation percentile because raw-citation counts have a time trend, as older publications have had longer to accumulate citations. The use of percentile figures allows for like-to-like comparison of citations over time.

Interdisciplinary Research

WOS categories are assigned to every publication by subject-matter experts at Clarivate. Publications assigned multiple categories (e.g., thermodynamics *and* electrochemistry) are defined here as *interdisciplinary*. During each of the last five years, aerospace publications written by teams composed of researchers from the United States and China were assigned more WOS categories than the average aerospace publication (see Figure 4.7).[16]

Conclusion

This chapter has provided evidence that assessment of the net effect of scientific collaboration with China on U.S. interests is not straightforward. On one hand, U.S.-based organizations are increasingly coauthoring publications with Chinese organizations that have ties with the PLA. Furthermore, some of these publications are focused on topics (e.g., hypersonic systems and targeting systems) of direct military relevance. On the other hand, aerospace publications written by teams of researchers based in the United States and China have greater influence and are more interdisciplinary. Furthermore, these collaborations might improve bilateral relations in a way that is not measured here. In sum, assessment of the net benefit of collaboration requires making judgments about the relative importance of these effects as well as identifying additional risks and benefits.

FIGURE 4.7

Average Number of WOS Categories Assigned to Publications, 2016–2020

SOURCE: Features WOS data (Clarivate, undated-b).

---

[16] Except for the 2017 difference, the interdisciplinary metrics are not statistically significant at the 0.05 level.

CHAPTER 5

Conclusion

In the previous chapters, we presented the findings and methods of our exploratory analysis into three types of knowledge flows between the United States and China. As independent investigations, they present distinct approaches to studying the international flow of knowledge and provide empirical findings to inform policy decisions. Future research could tie these investigations together and extend the analysis. The research in this report raises the following questions:

1. To what degree are international mobility and collaboration correlated, or even causal, between academic researchers in the United States and China?
2. Are there certain university pairs, or clusters, where mobility and collaboration occur most often?
3. Do benefits and risks change across different types of mobility and collaboration?
4. What are the major one-way flows of knowledge (e.g., are there areas of scientific of technology dependency between the United States and China)?

For now, this report simply presents the independent mapping of knowledge flows—without normative judgements, policy recommendations, or commentary on their relation to one another—as a starting point to answering questions such as these.

Abbreviations

BIT	Beijing Institute of Technology
CAS	Chinese Academy of Sciences
DII	Derwent Innovation Index
HIT	Harbin Institute of Technology
MIT	Massachusetts Institute of Technology
PLA	People's Liberation Army
PRC	People's Republic of China
PRC10	Top 10 ranked People's Republic of China universities
SSons	Seven Sons of National Defense
S&T	science and technology
UC	University of California
US10	Top 10 ranked United States universities
WOS	Web of Science

References

Allison, Graham, Kevin Klyman, Karina Barbesino, and Hugo Yen, *Avoiding Great Power War Project: The Great Tech Rivalry: China vs the US*, Harvard Kennedy School Belfer Center for International Affairs, 2021.

Australian Strategic Policy Institute International Cyber Policy Centre, "China Defence Universities Tracker: Beihang University," webpage, May 5, 2021. As of September 1, 2022: https://unitracker.aspi.org.au/universities/beihang-university

Baas, Jeroen, Michiel Schotten, Andrew Plume, Grégoire Côté, and Reza Karimi, "Scopus as a Curated, High-Quality Bibliometric Data Source for Academic Research in Quantitative Science Studies," *Quantitative Science Studies*, Vol. 1, No. 1, Winter 2020.

Cao, Cong, Jeroen Baas, Caroline S. Wagner, and Koen Jonkers, "Returning Scientists and the Emergence of China's Science System," *Science and Public Policy*, Vol. 47, No. 2, April 2020.

Clarivate, "Derwent Innovations Index on Web of Science," database, undated-a. As of January 19, 2022: https://clarivate.com/webofsciencegroup/solutions/webofscience-derwent-innovation-index/

Clarivate, "Web of Science: Search," webpage, undated-b. As of January 19, 2022: https://www.webofscience.com/wos/woscc/basic-search

Elsevier, "Physical Sciences and Engineering," webpage, undated-a. As of March 15, 2023: https://www.elsevier.com/physical-sciences-and-engineering

Elsevier, "Scopus: Affiliations," webpage, undated-b. As of March 15, 2023: https://www.scopus.com/search/form.uri?display=basic#affiliation

Elsevier, "Scopus: Search for an Author Profile," webpage, undated-c. As of March 15, 2023: https://www.scopus.com/freelookup/form/author.uri

Elsevier, *Scopus Content Coverage Guide*, 2020.

Fedasiuk, Ryan and Emily Weinstein, "Overseas Professionals and Technology Transfer to China," *Issue Brief*, Center for Security and Emerging Technology, July 21, 2020. As of September 6, 2022: https://cset.georgetown.edu/publication/overseas-professionals-and-technology-transfer-to-china/

Fedasiuk, Ryan, Emily Weinstein, and Anna Puglisi, *China's Foreign Technology Wish List*, brief, Center for Security and Emerging Technology, May 2021.

Guerrero Bote, Vicente P., Carlos Olmeda-Gómez, and Félix de Moya-Anegón, "Quantifying the Benefits of International Scientific Collaboration," *Journal of the American Society for Information Science and Technology*, Vol. 64, No. 2, 2013.

Gong, Hong, Libing Nie, Yuyao Peng, Shan Peng, and Yushan Liu, "The Innovation Value Chain of Patents: Breakthrough in the Patent Commercialization Trap in Chinese Universities," *PLoS ONE*, Vol. 15, No. 3, 2020.

Google Patents, "General Advanced Upper Stage of Solid Launch Vehicle," webpage, undated. As of March 15, 2023: https://patents.google.com/patent/CN105841556B/en

Hottenrott, Hanna, Michael E. Rose, and Cornella Lawson, "The Rise of Multiple Institutional Affiliations in Academia," *Journal of the Association for Information Science and Technology*, Vol. 72, No. 8, August 2021.

Hottes, Alison K., Marjory S. Blumenthal, Jared Mondschein, Matthew Sargent, and Caroline Wesson, *International Basic Research Collaboration at the U.S. Department of Defense: An Overview*, RAND Corporation, RR-A1579-1, 2023. As of March 15, 2023: https://www.rand.org/pubs/research_reports/RRA1579-1.html

Institute of International Education, "All Places of Origin," spreadsheet, 2022. As of November 15, 2022: https://opendoorsdata.org/data/international-students/all-places-of-origin/

Joske, Alex, *Picking Flowers, Making Honey*, Australian Strategic Policy Institute, 2018.

Joske, Alex, *The China Defence Universities Tracker: Exploring the Military and Security Links of China's Universities*, Australian Strategic Policy Institute, November 25, 2019.

Kynge, James, "China's High-Tech Rise Sharpens Rivalry with the US," *Financial Times*, January 19, 2022.

Laudel, Grit, "Studying the Brain Drain: Can Bibliometric Methods Help?" *Scientometrics*, Vol. 57, No. 2, 2003.

Ledford, Heidi, "How to Solve the World's Biggest Problems," *Nature,* Vol. 525, No. 7569, 2015.

Lloyd-Damnjanovic, Anastasya and Alexander Bowe, *Overseas Chinese Students and Scholars in China's Drive for Innovation*, U.S. China Economic and Security Review Commission, October 7, 2020. As of November 15, 2022: https://www.uscc.gov/sites/default/files/2020-10/Overseas_Chinese_Students_and_Scholars_in_Chinas_Drive_for_Innovation.pdf

Moed, Henk F., M'hamed Aisati, and Andrew Plume, "Studying Scientific Migration in Scopus," *Scientometrics*, Vol. 94, No. 3, 2013.

Mulvenon, James, and Chenny Zhang, "Targeting Defense Technologies," in William C. Hannas and Didi Kirsten Tatlow, eds., *China's Quest for Foreign Technology: Beyond Espionage*, Routledge, 2020.

Popper, Steven W., Marjory S. Blumenthal, Eugeniu Han, Sale Lilly, Lyle J. Morris, Caroline S. Wagner, Christopher A. Eusebi, Brian Carlson, and Alice Shih, *China's Propensity for Innovation in the 21st Century: Identifying Indicators of Future Outcomes*, RAND Corporation, RR-A208-1, 2020. As of March 15, 2023: https://www.rand.org/pubs/research_reports/RRA208-1.html

Portman, Rob, and T. Carper, *Threats to the U.S. Research Enterprise: China's Talent Recruitment Plans*, staff report to the Permanent Subcommittee on Investigations, U.S. Senate, November 18, 2019.

Public Law 115-232, John S. McCain National Defense Authorization Act for Fiscal Year 2019; Section 1286, Initiative to Support Protection of National Security Academic Researchers from Undue Influence and Other Security Threats, August 13, 2018.

Public Law 116-283, William M. (Mac) Thornberry National Defense Authorization Act for Fiscal Year 2021; Section 223, Disclosure of Funding Sources in Applications for Federal Research and Development Awards, January 1, 2021.

Robinson-Garcia, Nicolás, Cassidy R. Sugimoto, Dakota Murray, Alfredo Yegros-Yegros, Vincent Larivière, and Rodrigo Costas, "The Many Faces of Mobility: Using Bibliometric Data to Measure the Movement of Scientists," *Journal of Informetrics*, Vol. 13, No. 1, 2019.

Schmid, Jon, "The Determinants of Military Technology Innovation and Diffusion," dissertation, Georgia Institute of Technology, 2018.

Schmid, Jon, "Technological Emergence and Military Technology Innovation," *Defence and Peace Economics*, June 2022.

Schmid, Jon, and Fei-Ling Wang, "Beyond National Innovation Systems: Incentives and China's Innovation Performance," *Journal of Contemporary China*, Vol. 26, No. 104, 2017.

Shanghai Ranking, "2022 Academic Ranking of World Universities," webpage, 2022. As of April 21, 2023:
https://www.shanghairanking.com/rankings/arwu/2022

Turekian, Vaughan, "The Evolution of Science Diplomacy," *Global Policy*, Vol. 9, No. 53, 2018.

U.S. Department of Defense, *DoD International Science and Technology Engagement Strategy*, 2020.

U.S. Department of Energy Order 486.1, "Department of Energy Foreign Government Talent Recruitment Programs," June 7, 2019.

U.S. Department of Energy Order 142.3A, "Unclassified Foreign Visits and Assignments Program," October 14, 2010.

U.S. Department of Energy Order 142.3B, "Unclassified Foreign National Access Program," January 15, 2021.

U.S. Department of Justice, "Information about the Department of Justice's China Initiative and a Compilation of China-Related Prosecutions Since 2018," November 19, 2021. As of April 20, 2023:
https://www.justice.gov/archives/nsd/information-about-department-justice-s-china-initiative-and-compilation-china-related

U.S. Government Accountability Office, *U.S. Universities in China Emphasize Academic Freedom but Face Internet Censorship and Other Challenges*, U.S. Government Printing Office, GAO-11-502, June 2011. As of November 15, 2022:
https://www.gao.gov/products/gao-16-757

Wagner, Caroline S., and Xiaojing Cai, "Drop in China-USA International Collaboration," *ISSI Newsletter*, Vol. 18, No. 1, March 2022.

White House, Presidential Memorandum on United States Government-Supported Research and Development National Security Policy: National Security Presidential Memorandum – 33," January 14, 2021. As of April 20, 2023:
https://trumpwhitehouse.archives.gov/presidential-actions/presidential-memorandum-united-states-government-supported-research-development-national-security-policy/

White House, "Suspension of Entry as Nonimmigrants of Certain Students and Researchers from the People's Republic of China," Proclamation 10043, *Federal Register*, Vol. 85, No. 34353, May 29, 2020.

Zwetsloot, Remco, Emily Weinstein, and Ryan Fedasiuk, *Assessing the Scope of U.S. Visa Restrictions on Chinese Students*, Center for Security and Emerging Technology, February 2021.